SUMMER
IN THE WOODS

written and photographed
by
Mia Coulton

In the summer,

the days can be very hot.

Take a walk in the woods.
The trees have lots of green leaves
that make the woods
shady and cool.

On the walk, look for **insects**.

An insect has three body parts and six legs.

Look closely down on the ground.

Look, it's an **ant**.

An ant is an insect.

Look for insects

near **shallow** water in the woods.

Look closely on the leaves.

A fly is resting on a leaf.

A fly is an insect.

A dragonfly is resting on a leaf.

A dragonfly is an insect.

Look for insects

at the edge of the woods.

Look closely in the flowers.

Look, it's a bee.

It's a **honeybee**!

Look, it's a yellow bee with black stripes.

It's a **bumblebee**!

There are many insects to look for in and around the woods in the summer.

Chirp! Chirp!

Look, it's a **grasshopper!**

Glossary

ant: A small insect that lives in large groups called colonies

bumblebee: A large, hairy bee that makes a loud buzzing sound when it flies

dragonfly: An insect with two long, thin pairs of outstretched wings

fly: A type of flying insect with two wings

grasshopper: A chirping insect that has long hind legs that it uses for jumping

honeybee: A bee that makes honey from flower nectar

insect: A small animal with three body parts and six legs

shallow: In water, an area that is not very deep